Bibliografische Information der Deutschen Nationalbibliothek:

Die Deutsche Bibliothek verzeichnet diese Publikation in der Deutschen National-
bibliografie; detaillierte bibliografische Daten sind im Internet über http://dnb.d-
nb.de/ abrufbar.

Impressum:

Copyright © 2006 GRIN Verlag, Open Publishing GmbH
Druck und Bindung: Books on Demand GmbH, Norderstedt Germany
ISBN: 9783640376926

Yvonne Papadopoulos

Testing a hypothesis with the methods of conceptual biology

A literature-based approach

GRIN Verlag

Testing a hypothesis
with the methods
of conceptual biology

Bachelor Thesis
Study course Molecular Biology
Main focus on bioinformatics

Presented
by
Yvonne Papadopoulos
the
08/21/2006

This project was realized at

The Center for the Biology of Natural Systems

Queens College, City University of New York

Fachhochschule Gelsenkirchen

Fachbereich Angewandte Naturwissenschaften

August-Schmidt-Ring 10

45665 Recklinghausen

If I speak in human and angelic tongues but do not have love, I am a resounding gong or a clashing cymbal. And if I have the gift of prophesy and comprehend all mysteries and all knowledge; if I have all faith so as to move mountains but do not have love, I am nothing. If I give away everything I own, and if I hand my body over so that I may boast but do not have love, I gain nothing. Love is patient, love is kind. It is not jealous, [love] is not pompous, it is not inflated, ⁵it is not rude, it does not seek its own interests, it is not quick-tempered, it does not brood over injury, it does not rejoice over wrongdoing but rejoices with the truth. It bears all things, believes in all things, hopes all things, endures all things. Love never fails. If there are prophecies, they will be brought to nothing; if tounges, they will cease; if knowledge, it will be brought to nothing. For we know partially and we prophesy partially, but when the perfect comes, the partial will pass away. When I was a child, I used to talk as a child, think as a child, reason as a child; when I became a man, I put aside childish things. At present, we see indistinctly, as in a mirror, but then face to face. At present I know partially; then I shall know fully, as I am fully known. So faith, hope, love remain, these three; but the greatest of these is love.

- 1. Corinthians, 1-13 -

This thesis is dedicated to Michaela Papadopoulos, who passed after a long fight of hope and despair in February 2006. Though not been in harmony through all the years of relationship, we shared something unmeasurable: faith, hope and love. The faith in god, the hope for a better future and the love for the god in ourselves. I will never disappoint the confidence you gave me and will raise your children as they were mine. And moreover I am grateful to be the chosen one. I hope you are fine now.

Table Of Contents

1. Abstract

Undoubtedly, the development of Conceptual Biology, which uses text-mining applications specific to biology is the only way to cope with the increasing amount of free textual data produced in this field. The increasing interest of users in efficiently retrieving and extracting relevant information, the need to keep up with new discoveries described in the literature or in biological databases, and the demands posed by the analysis of high throughput experiments, are the underlying forces motivating the development of Conceptual Biology tools, such as text-mining applications in molecular biology. Therefore the methods of Conceptual Biology have been used for this study to test the hypothesis, that genetically modified foods have no impact on public health. We studied the records of databases and those ones of Literature Based Discovery tools. After the binary scoring of the records with respect to their usefulness, they were also classified by their positive, neutral or negative conclusions with respect to the effect of genetically modified food on public health. In conclusion, we have to deny the hypothesis and therefore to state that genetically modified foods have an impact on public health. Further studies in conceptual biology may focus what kind of impact genetically modified food has on public health.

2. Introduction

"Knowledge can be created by drawing inference from what is already known."

-Davies, R. 1989-

The abundance of electronically accessible texts is rising exponentially throughout the last decade. A vast amount of digital information – especially in molecular biology and genetics- is seemingly an auspicious resource for conceptual biology. In the demands of biomedical or biochemical investigators for sources or references, librarians and information specialist are commonly puzzled.

The increasing amount of scientific journals, with an even greater number of articles per journal, expands already in humongous bibliographic databases. The rapid and persistent augmentation in the number of biological, biomedical and even genetics publications gives rise to the desperate situation, that researchers can no longer read more than minute proportion of the literature in their field. Dealing with the substantial quantity of information has induced a fragmentation of scientific literature (Ganiz et al. 2005), that exists within:

1. *specialities*: advances in the research field e.g. modern comforts in biophysics or mathematical physics
2. *sub-specialiti*es: subordinated field in the research field e.g. proteomics or aquatic toxicology
3. *structure*: structure that occurs in the research field e.g. blood, cell or lipid
4. *technique*: special techniques that can be found in the research field e.g. mass spectrometry or gel electrophoresis

This specialisation or fragmentation of scientific literature leads to an insuperable border and furthermore to an increasing problem in science, particularly with regard to biomedicine (Swanson 2001).

Scientists incline the correspondence more within their fragments than with the scientific field´s farther community, enhancing the lack of communication between specialities (Swanson et al. 1997). This argument is proven within the citations of literature of authors, that cite heavily those of their own narrow specialities.

Thus, scientists may never be aware of the published data and results of others´ relevant work. In addition, this gives rise to useful and important connections between fragmented and implicit data, but yet unnoticed.

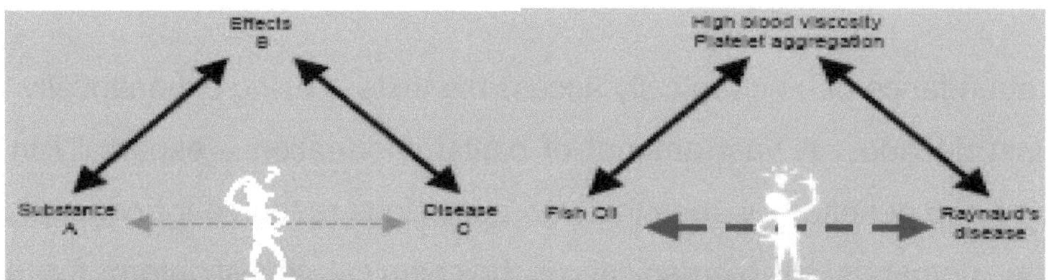

Figure 1. Swanson´s discovery. The relationships AB and BC are known and reported in the literature. The implicit relationship AC is a putative new discovery (Weeber et al. 2001)[1] Swanson´s serendipitous literature-based discovery of a cure for Raynaud´s didease by dietary fish oil. The literature for both issues were disjoint. If these scientific fields had been aware of each other, the cure would have been found much earlier than Swanson´s discovery

For this reason, conceptual biology allows one to encompass, without limitation as many fields of science as necessary. This is important because one cannot work experimentally in every scientific field. It is predicted that the most important changes in cellular and molecular biology will be conceptual. In turn it will be conceptual biology, supporting a need for data collection and phenomenological publications, but, and that is the most important thing, how these collections and publications are connected and related.

Hence, Informational Retrieval and the conventional computer-aided literature searching represent insufficient techniques for recognizing useful connections. Thus, this leads to Literature Based Discovery (LBD), that directly addresses the limits of knowledge discovery. LBD is a tool, that gives rise to occurrence of novel connections, that have yet not been published.

The concept´s principle is based on the hypothesis that *"wealth of recorded knowledge is greater than the sum of its parts."* (Davies 1989). As a precursor in LBD, Don R. Swanson introduced in 1986 his concept of discovering new associations within a bibliographic database. Furthermore he has asserted hypothesis that have been published in various articles (Swanson 1986; 1988; 1990; Smalheiser & Swanson 1996, 1998). According to his statement LBD is a process, finding complementary structures in apparently disjoint literature.

---

[1] In different works disease can be defined as A and substance can be defined as C. Thus, the search can either start from disease or substance.

These complementary structures derive from two discrete arguments, yielding novel and important inferences and insights, when combined. Those discrete arguments, that do not mention, cite or co-cite each other, are defined as "disjoint" arguments (Fig.1, p.7)

To specify the main purpose of LBD, I will focus the relevant traits:

1. LBD avails present knowledge from published science literature (e.g. Medline)

2. LBD is a process that strives to find relations between two disjoint arguments (e.g." high blood viscosity" and "platelet aggregation" are mentioned arguments in separate literature of Fish Oil and Raynaud´s Disease)

3. The combination of these arguments may obtain a new non obvious insight

4. Any connection made should be novel and previously unpublished (e.g. no publication ever mentioned Fish Oil and Raynaud´s Disease together)

An intricacy of LBD, because it comprises two types of entities: concept and literature.

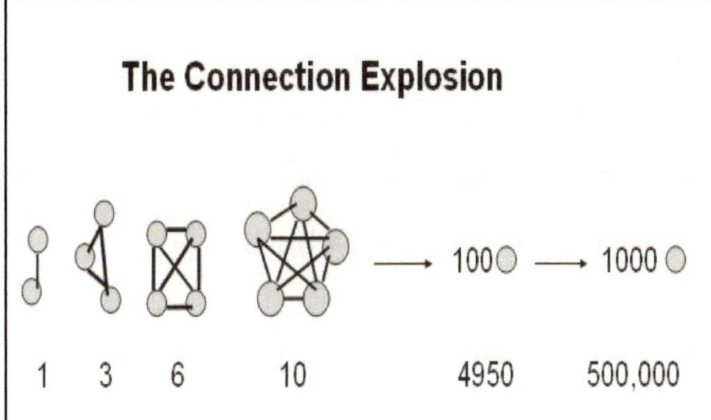

Figure 2. The Connection Explosion. To emphasize the problem, take a look at the number of potential connections between units of specialized literatures, that grows much faster than the number of units themselves. The number of pairwise connections increases with the formula $x = [n(n-1)/2]$.
Hence, we would receive for 10^7 *Medline* records 50,000 billion possible 2-way connections between individual articles.

But there are three more serious obstacles for LBD. First, there is seeminlgy unmanageable information space with many potential relations due to the vast amount of data and text. Second, the language itself represents in an unstructured format with characteristic grammar and semantic – even more there embedded in different languages.

Third, the lack of standardized vocabulary inhibits the process to formally define various LBD techniques. Swanson´s example reflects, that we could discover new knowledge from available existing text, if we can assemble the pieces of existing knowledge in the right way.

Conceptual Biology is not a common term in the field of biochemical and biomedical research. Indeed, it leads to important information, enveloped in the overwhelming multitude of literature. Thus, researchers could pose the question of a comprehensible definition of what conceptual biology is really about. Mikhail Blagosklonny hit the bull´s eye, when he says, that a conceptual biologist *"...can generate a hypothesis in which predictions are formulated in testable terms, and then search for relevant information among published reports of experiments that may have had a different purpose altogether."* (Blagosklonny et al. 2002).

In the common field of biochemical investigations, *"one is not licensed to theorize without providing new data"* (Blagosklonny et al. 2002), but according to D. Bray (Bray 2001) this *"is a sociological problem and not a scientific one."* Furthermore conceptual biology is an important and irreplaceable complement to the accepted empirical biology in part, because researchers struggle to maintain expertise and management in their fields and even more to understand the connections between different research fields that could reveal fundamental new facts, embedded in the overproduced data field.

Hypothesis testing is central to the process of scientific discovery, and experimental design is a common methodology in the evidence gathering part of hypothesis testing. But some sort of what is commonly now known as 'data mining' methodologies can and always have been used too for the data gathering part of hypothesis testing. However, and maybe even ironically, data mining recently has been confined by the vast expansion of knowledge that is increasingly stored in specialized databases and formats as will be explained bellow. To avert these limitations, new developments have emerged in the form of what has been called Conceptual Biology.

With specialized tools and methodologies which are also explained below, Conceptual Biology allows seemingly unlikely hypothesis testing to be performed, examples of which will be given bellow also.

3. The scientific problem

As the scientific environment for this research project was to work in the Critical Genome Project, the purpose of my assay was to conduct research on the scientific basis of genetic engineering and related aspects of biotechnology with the methodology that is embodied within Conceptual Biology. The special area of interest was the impact of genetically modified food on public health and the attendant scientific problem was to test the hypothesis: Genetically modified food has no impact on public health!

To confirm or deny the hypothesis, we had to test it with formulated and testable predictions against published literature, by recognising, establishing and exploiting the links between seemingly unrelated topics, using databases and the tools of Literature based Discovery.

4. Materials and Methods

Materials and methods are sorted like the way they have been used during the process of the hypothesis testing.

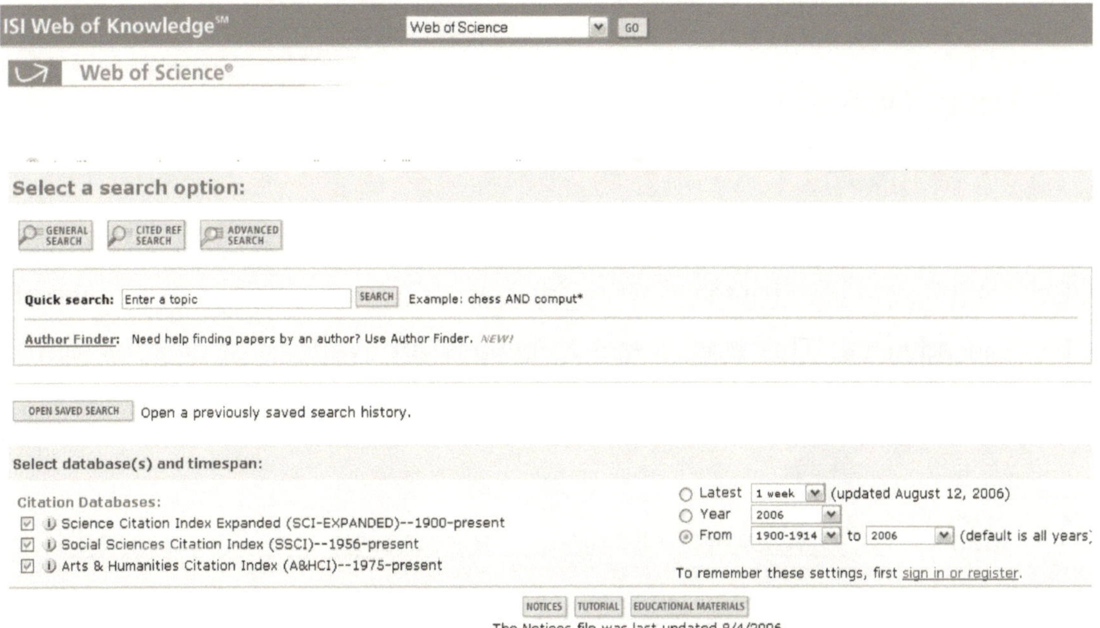

Figure 3. Web of Science. The homepage presents three different search options. The general search is a search for records by topic, author name, source title, and author address. With the Cited Reference Search the user can search for articles that cite other works that you select from the citation index. By using the advanced search the user can create complex searches using field tags and set combinations. For the goal of this thesis we choosed the general and the advanced search modus.

4.1 MeSH (http://www.ncbi.nlm.nih.gov/entrez/query.fcgi?db=mesh)

PubMed uses a controlled vocabulary to index the articles in the database. This controlled vocabulary is called Medical Subject Headings (MeSH). Each citation in PubMed is assigned a series of subjects, or *MeSH* Headings, to identify the topics covered in an article.

MeSH provides consistent way to retrieve information that may use different terminology for the same concepts. When doing a keyword search, the user may miss key articles. If the exact keyword is not used, *PubMed* may not retrieve that article. Not every concept has a corresponding *MeSH* term, but it is always a good idea to search *MeSH* before doing a basic keyword search.

We started the *MeSH* term database search with the words "Genetically modified food" and its suggestions. Furthermore we searched for the term "Public Health" to finally combine these two terms by the Boolean operator AND.

Those terms then have been send to the *Pubmed* database. Moreover, all the offered *MeSH* terms and its combinations were used for the search in the following databases.

4.2 Using databases

With total access to all databases listed in the Levy Library of the Mount Sinai School (http://fusion.mssm.edu/levy/databases/), we decided to use only eight of them in advance. This was meant to bypass an overload of digital information and paper. So the used databases had been:

1. Medline (http://www.ncbi.nlm.nih.gov/entrez/query.fcgi?DB=pubmed)

2. Web of Science (http://scientific.thomson.com/products/wos/)

3. Faculty of 1000 Biology (http://www.f1000biology.com/start.asp)

4. Faculty of 1000 Medicine (http://www.f1000medicine.com/about/biology)

5. ProQuest Digital Dissertations (http://wwwlib.umi.com/dissertations/)

6. Wiley Interscience (Online Books) (http://www3.interscience.wiley.com/)

7. BioOne (www.bioone.org)

8. Google Scholar (www.scholar.google.com)

Medline (Medical Literature Analysis and Retrieval System Online) is an international literature database, offering digital essays of life sciences and biomedical information, covering the fields of medicine, nursing, pharmacy, dentistry, veterinary medicine, and health care. In addition, the *Medline* database covers most of the literature in biology and biochemistry and contains more than 13 million records from nearly 4,800 selected publications covering biomedicine and health from 1966 to the present. The database is freely accessible via the *PubMed* interface, and new citations are added Tuesday through Saturday[2]

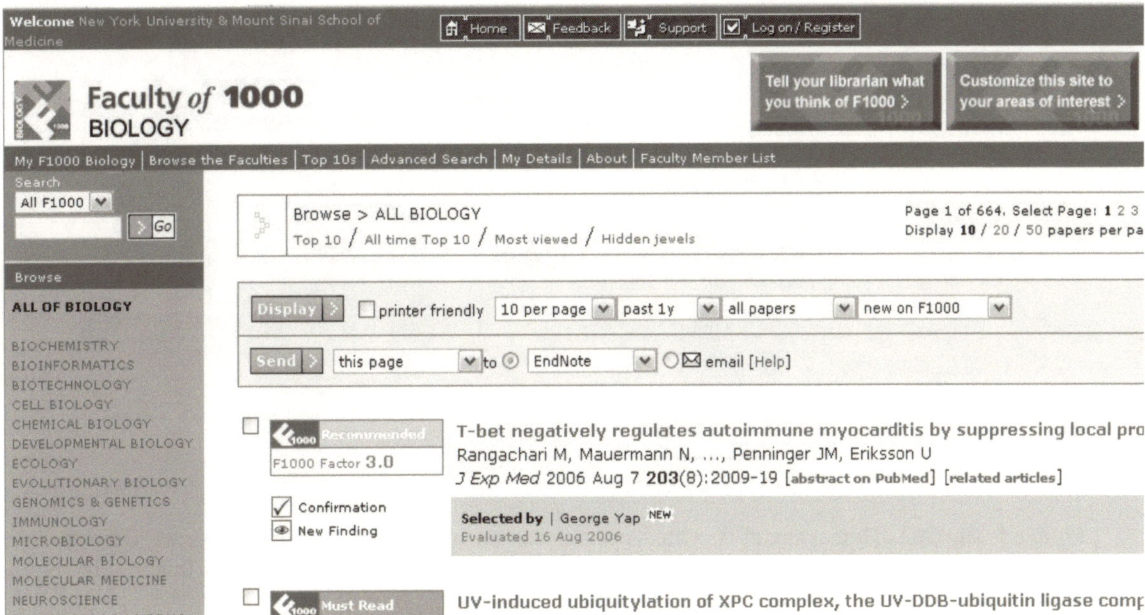

Figure 4. Faculty of 1000 biology. Homepage of the research database launched by the Biomed Central Ltd. (**http://www.f1000biology.com/start.asp**), received 08-16-2006

The *Web of Science* (Fig.3, p.10) is an integrated, versatile research database, providing easy access to high quality, diversified scholarly information in the sciences, social sciences, and arts and humanities, as well as search and analysis tools that enhance this content. Within the *Web of Science*, users can

---

[2] http://www.ncbi.nlm.nih.gov/entrez/query.fcgi?DB=pubmed , received 08-16-2006

search for required information in international journals, open access resources, books, patents, proceedings, or Web sites. *Web of Science* welcomes the competent user with more than 16 million full-text links to 10,000 journals from more than 300 publishers[3].

Faculty of 1000 Biology (Fig.4, p.12) is a new online research tool that highlights the most interesting papers in biology, based on the recommendations of over 1000 leading scientists. *Faculty of 1000 Medicine* belongs like the *Faculty of 1000 Biology* to the Biomed Central Databases, that delivers a collection of free biomedical, medicine, health, biotechnology related online journal articles, and provides a consensus view of distinguished articles referred to the trends in medicine. *Faculty of 1000 Medicine* offers similar functionality and organisation to its sister service, the highly successful *Faculty of 1000 Biology*, to which more than 80% of the world's top institutions subscribe[4].

Figure 5. BioOne.. Homepage of BioOne. A search can be arranged by the Title, Abstract, Descriptors (controlled index terms), Identifiers (uncontrolled index terms) and Author Keywords fields.

ProQuest Digital Dissertations enables the user to freely access the most current two years of citations and abstracts in the Dissertation Abstracts database. The entire database consists of more than 1.6 million titles of digital dissertations and thesis.

---

[3] http://scientific.thomson.com/products/wos/ , received 08-16-2006
[4] http://databases.biomedcentral.com/, received 08-16-2006

Wiley InterScience was introduced in 1997 and launched commercially in January 1999. As a leading international source for quality content promoting discovery across the spectrum of scientific, technical, medical and professional endeavors, *Wiley InterScience* has built its prestige by continuously appending novel content up to more than 2,500 journals, books, reference works, databases, laboratory manuals and *The Cochrane Library*, which is the world's best-known resource for evidence-based medicine[5].

BioOne (Fig.5, p.13) is a database of high-impact, peer-reviewed bioscience research journals from non-profit scholarly publishers. As a collaboration of scientific societies, libraries, academe, and the commercial sector, *BioOne* provides access to a thoroughly linked information resource of interrelated journals, focused on the biological, ecological, and environmental sciences[6].

Google Scholar Search Engine enables you to search specifically for scholarly literature, including peer-reviewed papers, theses, books, reprints, abstracts and technical reports from all broad areas of research.

4.3 Using Literature Discovery Tools

"We think too small. Like the frog at the bottom of the well. He thinks the sky is only as big as the top of the well. If he surfaced, he would have an entirely different view."

- Mao Tse Tung[7] -

A program or approach is knowledge driven if it uses relatively general knowledge (including knowledge of how to search combinatorial spaces) as a main source of power. A data-driven program instead relies on specific measurements, statistics, or examples. Most approaches are some combination of both.

---

[5] http://www3.interscience.wiley.com/aboutus/, received 08-16-2006
[6] http://www.bioone.org/perlserv/?request=index-html, received 08-16-2006
[7] http://www.ils.unc.edu/~cablake/Papers/BlakeAnderson.pdf

A same task might be approached in either way: consider a chess program that bases its next move either on matching against a large database of examples of previous positions and good moves, or on general heuristics about strategy and tactics combined with a search over possible successive moves.

The input to a data-driven program can be megabytes of data, whereas the input to a knowledge driven program could be a mere few lines. Depending on the specifics of the problem, either might take longer to run.

4.3.1 Arrowsmith

The domain of *Arrowsmith* is still the medical literature and especially the Medline database, because it has been specifically designed for *Medline* with its huge domain-specific stop-word lists. Titles of the start literature, also *Medline´s MeSH* (Lowe et al. 1994) headings and subheadings are used in the present www version, are downloaded from *Medline*.

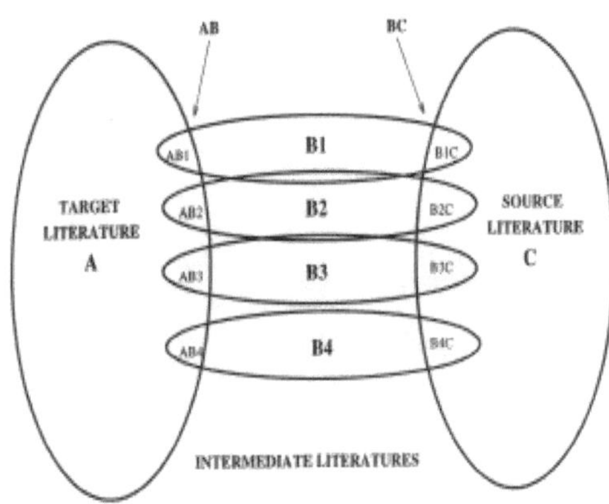

Figure 6. Venn diagram Of Swanson´s idea. Sets A and C are linked through intermediate sets Bi which contain the word Bi in their titles and which overlap both A and C. By examining the articles in the pairs of intersections ABi and BiC, useful information may be inferred regarding possible biological linkages among A, B and C. (A and C are shown here as having no articles in common. When there is overlap between sets A and C, the articles in the direct intersection should first be identified and evaluated prior to carrying out an *ARROWSMITH* search.) (Swanson et al.1998)

As prior work required manual selection of physiologically significant keywords, short phrases, and meaningful word combinations from each title, nowadays *Arrowsmith* is the automated version of this manual process. The extraction process still implicates excluding apparently unsuitable words (Swanson 1986), and promotes the following steps:

1. Exclude inappropriate words and array them in a pre-compiled exclusion list, or stop list to terms extracted from the titles. The list is human-compiled (on the basis of judgment applied *a priori* concerning the suitability of each word) and consists of more 5,000 words (Swanson & Smalheiser, 1997)

2. Relative frequency is used in the selection of the intermediate concepts (*B-concepts*). Each *B-concept* becomes the basis of a Medline query, and step 1 is applied to the resulting composite document set.

3. Identifying the terminals (target concepts of interest) is arranged by the number of links to intermediate topics, with topical limitations like "dietary," "toxicity," etc. The design of *Arrowsmith* tries to assist in the comprehension and recognition of intermediate connections between the *A* and *C* literature. The closed approach includes two Medline searches in the *Arrowsmith's* process; the first search defining the start literature (*A*) and the second defining the target literature (*C*).

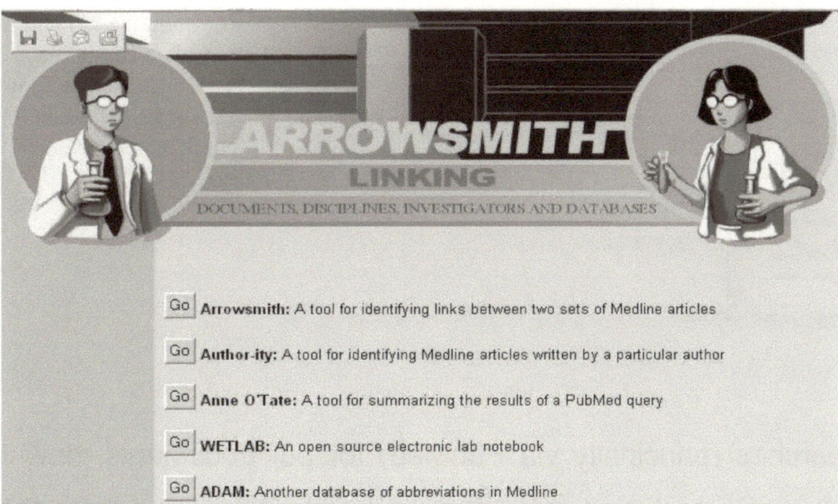

Figure 7. *ARROWSMITH* Homepage. The Homepage offers beside the B-term search tool *Arrowsmith* also different userfriendly specials to simplify the literature research workflow.(http://arrowsmith.psych.uic.edu/arrowsmith_uic/index.html), received 08-18-2006.

Then the program generates a list of words and phrases found in the titles of both literatures. To achieve sound B concepts, the generated list can be edited further through several steps by:

a) we selected only certain semantic categories: Disorders, Genes & Molecular Sequence, Living Beings, Phenomena and Physiology

b) we aligned a frequency threshold for B terms fewer than two times in the AB OR the AC literature

c) moreover we did a manual selection of concepts

4.3.2 BioRAT

BioRAT (http://bioinf.cs.ucl.ac.uk/biorat/) is an Information Extraction engine designed to enable researchers to find research papers, but can also read them itself and extract key facts from them for display.

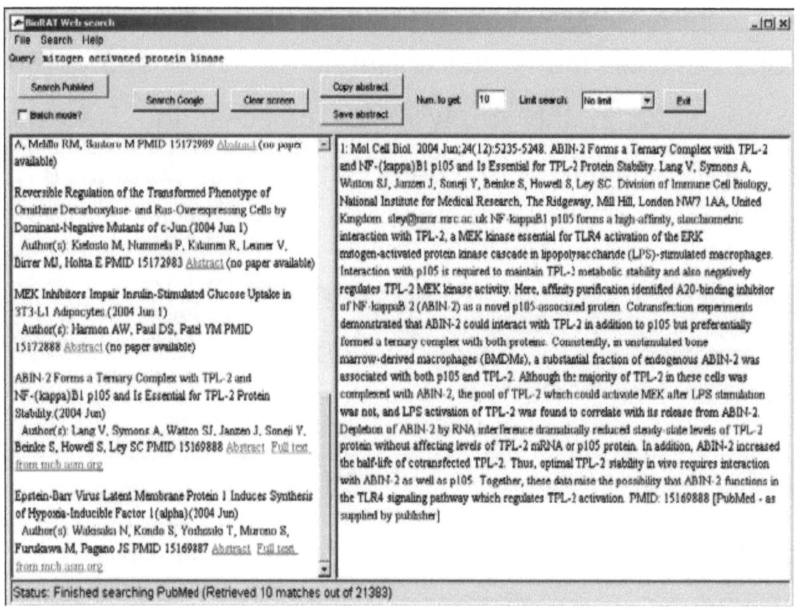

Figure 8. **Finding papers on the web.** (Corney et al. 2004) The document search interface. The user enters a query at the top and BioRAT accesses PubMed via the Internet. A list of matching titles (with date of publication, author, etc.) is shown on the left and the user can select any item to view the abstract, on the right or to download the full-length paper.

The interface searches (principally via PubMed) for pdf documents, downloads them, converts them to text and then applies Information Extraction patterns (using the GATE architecture [Cunningham et al 2002]) to find items of interest. *BioRAT* was designed to give people with no IE experience a high capacity tool, helping them to locate and analyse research papers. The system therefore combines tools to locate papers, to download full-length papers, to extract information from papers and to design templates to allow this extraction. Thus, the user enters a query into *BioRAT,* which is then passed on to *PubMed* for presenting a list of papers, from which abstracts or, where available, full-length papers can be chosen to download. After having obtained some text, the user is then moreover enabled to apply some pre-existing templates or create their own.

In either case, the templates match patterns in the text that contains potential information, which is extracted for display to the user and for feasible insertion into a database (Fig. 8)

A pronounced feature of *BioRAT* is the automatic location and acquirement of full-length papers wherever possible, instead of the commonly used abstracts.

The preferred method is to follow a series of hyperlinks to find each target paper. For the purpose of finding a particular paper, *BioRAT* starts with an URL (web address) provided by the *PubMed* database, enters the web page and identifies the corresponding hyperlinks and recursively follows links until it finds the target paper in PDF format.

BioRAT Options

Description	Value	Browse / Select
Proxy port number	none	...
PDF conversion script	D:\Yvonne\Eigene Dateien\Eigene Dateien\Programme\BioRat\biorat\scripts\dc_ps2ascii.bat	...
Gazetteer index file	D:\Yvonne\Eigene Dateien\Eigene Dateien\Programme\BioRat\biorat\data\gazetteer\lists2.def	...
USE_ABBREVIATIONS	yes	...
Template project file	D:\Yvonne\Eigene Dateien\Eigene Dateien\Programme\BioRat\biorat\data\templates\sample.xml	...
Results directory	D:\Yvonne\Eigene Dateien\Eigene Dateien\Programme\BioRat\biorat\data\results\	...
USE_RESIDUE_FINDER	no	...
Document store location	D:\Yvonne\Eigene Dateien\Eigene Dateien\Programme\BioRat\biorat\data\papers\	...
System files	D:\Yvonne\Eigene Dateien\Eigene Dateien\Programme\BioRat\biorat\etc\	...
GATE home directory	<unknown>	...
PDF Viewer	none	...
USE_STEMMER	yes	...
Temporary files location	D:\Yvonne\Eigene Dateien\Eigene Dateien\Programme\BioRat\biorat\tmp\	...
Proxy host name	none	...
Tokenizer script	D:\Yvonne\Eigene Dateien\Eigene Dateien\Programme\BioRat\biorat\scripts\biorattokeniser.rule	...
(Re-)create start-up scripts		Go!
Store these settings		Go!
Exit setup		Go!
Status:		

Figure 9. *BioRAT* Options and Settings. As the figure shows, abbreviations were embedded in the search. We did not use special gazetters, therefore the gazetter´s list was blank.

After being downloaded, the text is converted into a text-only version, ready for the IE engine. But in addition, the finding of a target paper is non-trivial for such a tool. The URL provided by *PubMed* (and ultimately, by the journal publishers) does not point to the paper itself, but rather to a web page from which the paper can be accessed.

We realized our search with both *BioRAT* Search Options, using the database of *Pubmed* and *Google*. No limit had been set on the search results and the number to get was chosen to add up to 200 records.

4.3.3 BITOLA

BITOLA uses HUGO (Human Genome Organisation), the National Center for Biotechnology Information's (NCBI) LocusLink (a database of curated sequence and descriptive information about genetic loci) and OMIM (NCBI's Online Mendelian Inheritance in Man catalog of human genes and genetic disorders) as sources for gene symbols and names as well as gene locations (http://www.mf.uni-lj.si/bitola/) .

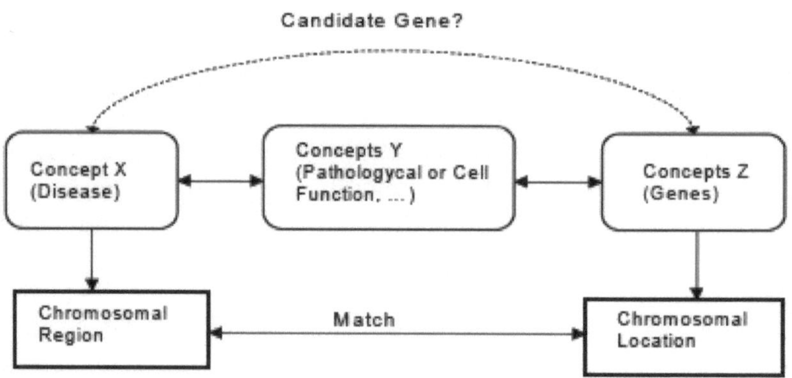

Figure 10. Discovery algorithm overview as applied to candidate gene discovery. For a starting disease *X*, we find the related concepts *Y* (disease characteristics) according to the literature (*Medline)*, then find the genes *Z* that are related to disease characteristics *Y*. (Hristovski et al., 2004)

Similar to Swanson's procedure, *BITOLA* first finds all the *concepts Y* that are related to the starting *concept X* (e.g., if *X* is a disease then *Y* might be pathological functions). Then all the *concepts Z* related to *concepts Y* are found (e.g., if *Y* is a pathological function, *Z* might be a molecule related to the pathophysiology of *Y*). Finally, the medical literature is searched to check whether *concept X* and *concepts Z* appear together. If they do not appear together, there is the possibility that a new relationship between *concept X* and *concept Z* has been discovered. Figure 10 illustrates the *BITOLA* literature discovery system.

BITOLA could not have been used in the first search runs, as there were not definite information about spezial modified genes and correlated to diseases in living beings in the beginning. Therefore this search engine was used in the later research to may deepen the information about modified genes, going along with diseases caused by the uptake of genetically modified food.

4.3.4 Manjal

The system *Manjal* ((http://sulu.info-science.uiowa.edu/Manjal.html) developed by Srinivasan consists of both an open and a closed discovery process. As usual, the open process starts with a search in *Medline* for documents about starting topic A. All *MeSH* terms extracted from this retrieved set R are candidate B-terms. These candidate B-terms are organized by *MeSH* semantic type. The list can be filtered by selecting semantic types. Only B-terms belonging to one or more of the selected types will be retained.

The B terms are ranked within each semantic type by weight and further filtered by retaining only the top B-terms. After searching the *Medline* for each of the left B-terms, *MeSH* terms are retrieved from each of the sets by selecting semantic types. Another organization by semantic type and another ranking by weight gives rise to a combination of lists, where the weight of each term is the sum of its weight in each of the separate lists. Hence, this combined list is refined by excluding all terms for which the *Medline* A-terms and the candidate C-terms returns non zero results. Finally, the open process´ result offers a list of C-terms organized by semantic type and ranked within each semantic type.

The closed process starts with two *Medline* searches for A-terms and C-terms. For both searches, the generated list of B-terms is organized by semantic types. After having been filtered by selecting semantic types, B-terms not belonging to one of these types are excluded. Thereupon the lists are merged. Only items appearing in both lists are retained by the weight according to the sum of their weights in the separate lists.

The resulting list is filtered by excluding all items for which a search in *Medline* for documents containing A, B, and C returns non zero results. Equally to the open process the result of the closed process is a list of B-terms organized by semantic type and ranked within each semantic type.

For the goal of this research we only used the closed process ability of the system according to a better comparison using the other LBD tools.

4.3.5 LitLinker

Like *BITOLA LitLinker* also uses *MeSH* terms to represent the documents. In the initial version of *LitLinker*, natural language processing methods to represent documents are used.

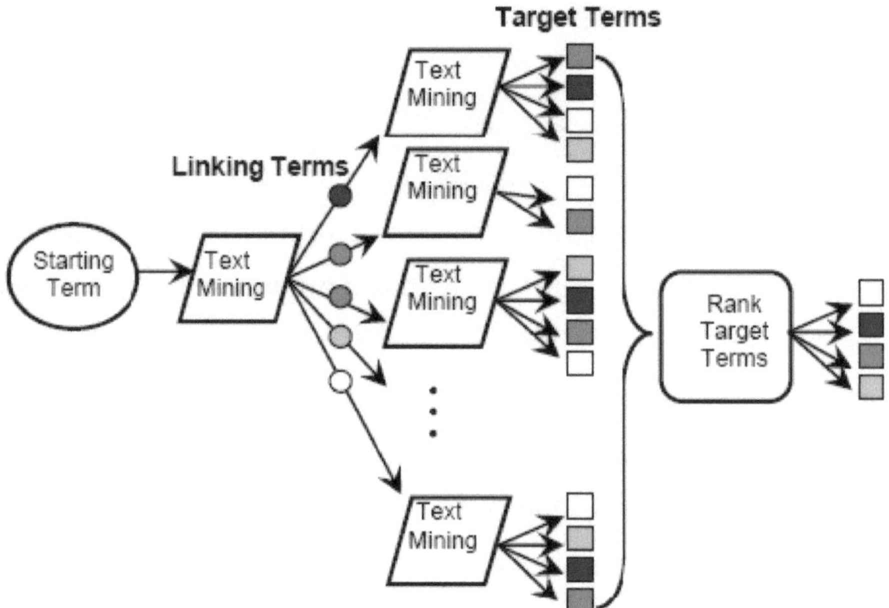

Figure 11. The discovery process in *LitLinker*. The literature-based discovery begins with a starting term, the term the researcher is interested in investigating. Next, *LitLinker* uses a text-mining process to find a set of terms that are directly correlated with the starting term. The inventers refer to this first set of correlated terms as the linking terms. For each of the linking terms, *LitLinker* then uses the same text-mining process to identify a set of terms that are correlated with each linking term. These final terms are the so called target terms. Finally, *LitLinker* ranks the target terms by the number of linking terms that connect the target term to the starting term. Thus, it provides an organized list of possibilities for this open-discovery process. (Pratt et al 2003)

The process begins with the researcher specifying a starting term and all documents containing that starting term are identified. Those documents are passed to a text mining component that identifies correlated terms in that literature. The first text-mining round generates the list of linking terms (shown as shaded circles). The process is then repeated for each of the linking terms to generate a set of target terms (shown as shaded squares). The target terms are pooled together and ranked based on the number of linking terms that generated the target term.

For both linking and target terms each color shade maps to a distinct term. LitLinker was designed with what Swanson calls an open discovery approach. A high-level view of the process is illustrated in Figure 11.

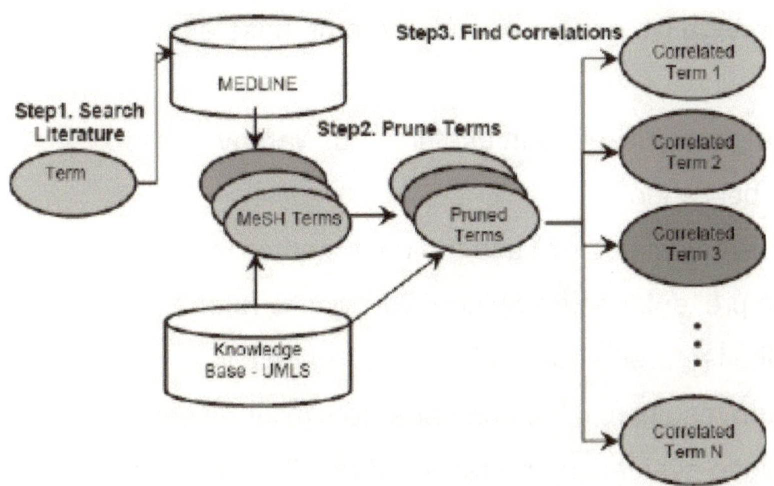

Figure 6. The text mining process in *LitLinker*. *LitLinker* uses a biomedical knowledge base, the Unified Medical Language System (UMLS), as an integral component throughout the text mining process[8]. This knowledge base was created by the National Library of Medicine (NLM) and contains over 975,000 biomedical concepts as well as 2.3 million concept names. The system was created by unifying hundreds of other medical knowledge bases and vocabularies to create an extensive resource that provides synonymy links as well as parent-child relation. (Pratt et al. 2003)

For the term that is provided, *LitLinker* identifies all documents in *Medline* that contain that term and gathers all the *MeSH* terms used in that collection of documents. For the literature search, an own local *Medline* database with the data leased from the NLM has been created. *LitLinker* searches this local database for collecting the literatures. Unlike the information retrieval tools

---

[8] National Library of Medicine. "Unified Medical Language System Fact Sheet" Available at http://www.nlm.nih.gov/pubs/factsheets/umls.html, received at 05/23/06

currently available to medical researchers, such as *PubMed, LitLinker* generates results about possible new connections between medical terms. *LitLinker* also provides an interactive web interface to display the identified correlations in an effective way (Skeels et al. 2005).

5. Evaluation and Interpretation

After gathering these links between seemingly unrelated fields and papers that either support the hypothesis or not, the most difficult barrier to overcome was to analyse each of the paper by an objective scoring at to whether or not it could be used in our hypothesis testing. Only than can an analysis be made to either reject the hypothesis or not. Therefore the papers were classified by giving each paper a positive or negative designation for each of these eight criteria:

(1)   Is the paper/dissertation useful to answer my question?
(2)   Who published?
(3)   Did the investigator(s) use the right method?
(4)   Is the presentation following the scientific rules?
(5)   Are the featured arguments of high grade?
(6)   Are the results reliable compared to similar dissertations?
(7)   Does the evaluation or analysis of results make sense?
(8)   Is the conclusion traceable?

According to the 1st criterion, useful (positive) papers included every topic on public health, in other words not only medical challenges, but also social and cultural alternations and risks that had been seen as an impact on public health. Topics having as a main issue the distribution of a new product and therefore of a new trademark or questions about insurances, labelling and legal questions were not the focus of interest here. Furthermore, if a company like Monsanto published a paper about one of its own genetically modified food product and maybe found no risk of this product to public health, this paper was useless (negative) to this thesis because of the conflicting interests of the researchers publishing it.

Therefore, for the 2nd criterion, it was always very important to ascertain, if the paper's publishers were dependant or independent researchers.

Next, and by way of an example, as some papers dealt with the digestion of genetically modified food, we gave a positive designation for the 3rd criterion only when the right methods of examinations were used, like the analysis of all digestive organs. In some papers, however, the most important organ, the pancreas and the analysis of its morphological alternation had neither been examined nor discussed. Furthermore, when a paper was following the right scientific rules, figures, charts and tables had been explained correctly, it also got a positive designation for the 4th criterion. Other papers for example, presented microscope pictures without any declaration of the magnification that had been used, therefore it was impossible to rely on the scientific discovery.

For the 5th criterion, useful papers had arguments of high grade and used current standards and devices, whereas other papers offered approximately indefinable morphological pictures caused by low light exposure. Papers deemed to be of main interest were listed and then their result was compared for the 6th criterion, although this classification tended to be more of a benchmark than a rating.

The next and 7th criterion was wether or not the correct evaluation of the received results was used. Therefore we checked if the right standard deviation was used to evaluate the results (empirical deviation of single observation or the standard deviation according to the true value of a quantity).

Finally we filtered the discussion and its context to the presented results, reviewing if the conclusion was traceable and sound according to the presented results, and we give a positive designation for when that was true.

As the eight categories had been phrased in a way to answer in a binary way, every positive answer was valuated with 1 point and every negative answer got no points, so that the highest possible score for any individual paper was 8 points. This objective classification of papers that could be used to test the hypothesis proved to be instrumental at this level. Only papers achieving the highest score were taken to the next level of evaluation and used in further analysis during this thesis

At the next level of evaluation, papers were classified as having concluded that genetically modified food have a positive, neutral, or negative effect on public health

6. Results

After the first search with the keywords "genetically modified food" AND "public health" *("Food, Genetically Modified"[MeSH] OR ("Food, Genetically Modified/adverse effects"[MeSH] OR "Food, Genetically Modified/economics"[MeSH]OR "Food, Genetically Modified/microbiology"[MeSH] OR "Food, Genetically Modified/poisoning"[MeSH] OR "Food, Genetically Modified/standards"[MeSH] OR "Food, Genetically Modified/statistics and numerical data"[MeSH] OR "Food, Genetically Modified/toxicity"[MeSH]) AND "Public Health"[MeSH])* we did not achieve records in every database or every discovery tool records. The databases *Faculty of 1000 Medicine* and *BioOne* did neither offer any record nor *ProQuest* and *Faculty of 1000 Biology*.

Therefore we changed the keywords to the *MeSH* terms *"genetically modified food"* AND " *Medicine"*. What happened next demonstrates the usefulness of the tools of conceptual biology, even though our first search gave us no matches.

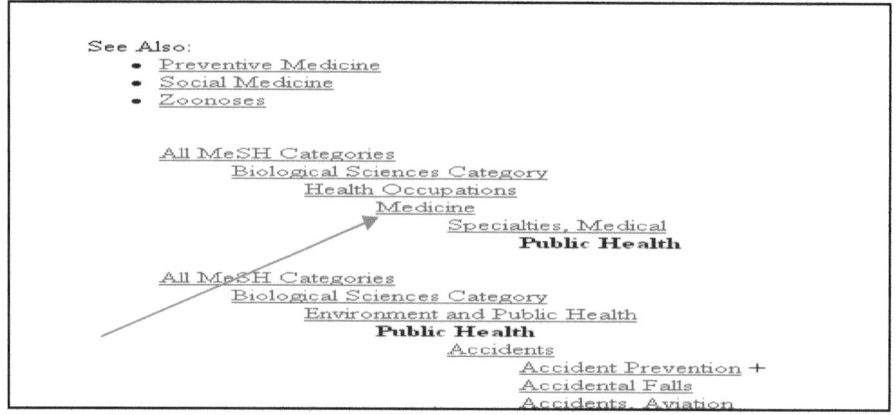

Figure 7. Exploding terms of public healths and superordinated headings. As we did not achieved records for the keyterms "genetically modified food" AND "public health" in every database, we changed the terms by the purpose of *MeSH* in the "genetically modified food" AND "medicine".

Using the new terms within the LBD tools, we achieved an number of new B-terms (or target concepts) illustrated in the following figure.

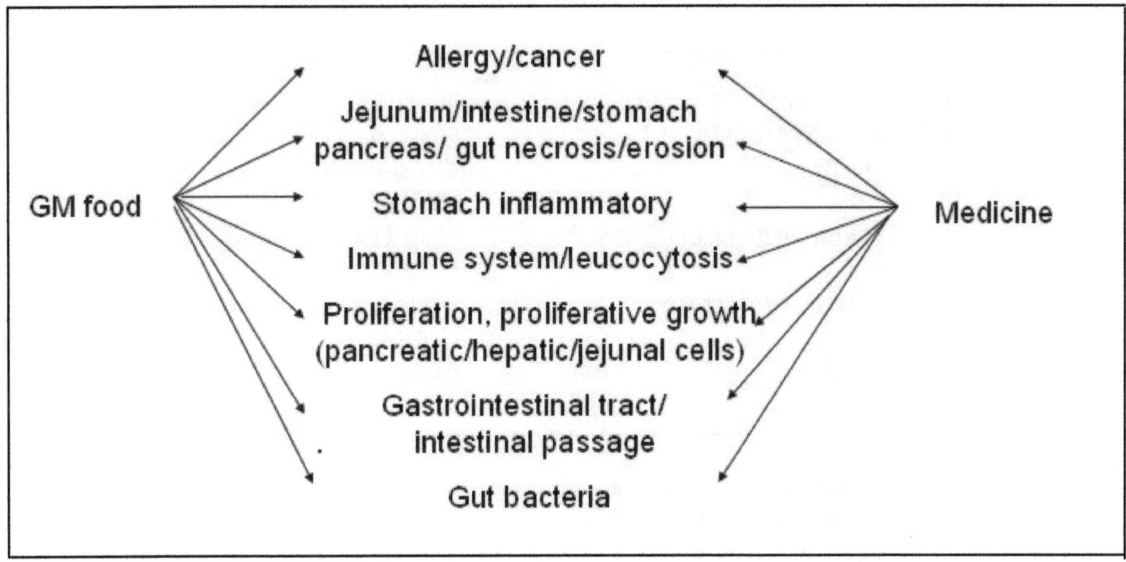

Figure 8. List of B-terms achieved by *Arrowsmith*, *Manjal* and *Litlinker*. The listed B-terms in the figure gave rise to deepen the search and to reveal unknown connections. GM food here is just taken as an abbreviation for the A-term "genetically modified food".

Just to get closer to new and unknown but related connections, we therefore combined the achieved list of new B-terms with the A-term *"genetically modified food"* and converted the achieved B-terms in C-terms for the following search process.

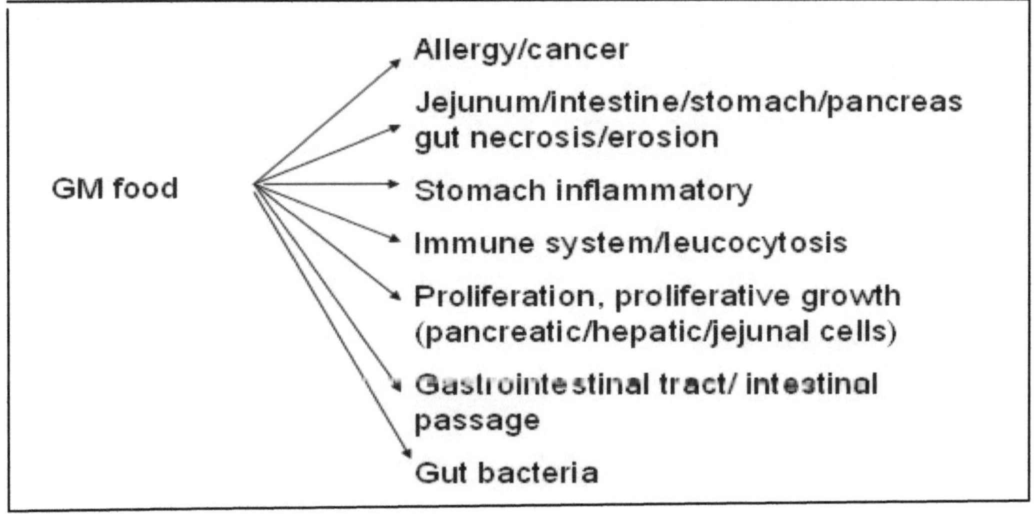

Figure 9. Combination of A-term with new C-terms. The achieved B-terms from the first search we converted in C-terms for the second search to obtain a secondary list of B-terms.

As a result we obtained a secondary list of B-terms by *Arrowsmith, Litlinker* and *Manjal.*

Flavr Savr™	Mon 810
New Leaf Potato ™	Phytat
Cry1Ab(c)	Golden rice
Bt	Vitamin A
maize/crop/toxin/potato	Hep B vaccines
/tomato	Glufosinat
Antibiotic resistance	Molecular Farming
gene	rBGH
gv-peas	fructane

Figure 10. Secondary list of B-terms. After having conducted a second search with the A-term "genetically modified food" combined with the C-terms shown in Fig.9. we obtained a secondary list of B-terms (or target concepts). The 21 terms listed here are just a part of 79 B-terms that occurred in all three closed approach tools.

In the first run with *Arowsmith*, the record of a B-term list was about 1409 papers. To limit the scope we restricted the list by semantic filters according to Chemicals & Drugs, Disorders, Genes & molecular Sequences & Protein Names, Living Beings, Phenomena and Physiology. The non highlighted B-terms were removed without having noticed the main focus of the publishers work. After restriction we still achieved a record of about 477 B-terms.

These B-terms were compared with the list of target concepts obtained by *LitLinker* and *Manjal.* Only those one, that occurred in every LBD tool (Fig. 8 and 9, p.26) had been taken for second search. After second search combining *"genetically modified food"* with the C-terms shown in Fig. 9, p. 26, we achieved the B-terms listed in Fig. 10. Again those B-terms were compared to the target concepts of *LitLinker.* Only those one occurring in both lists had been taken for further research on the papers.

The records of the first run with the terms [genetically modified food] AND [medicine] in the databases is presented in Fig. 11.

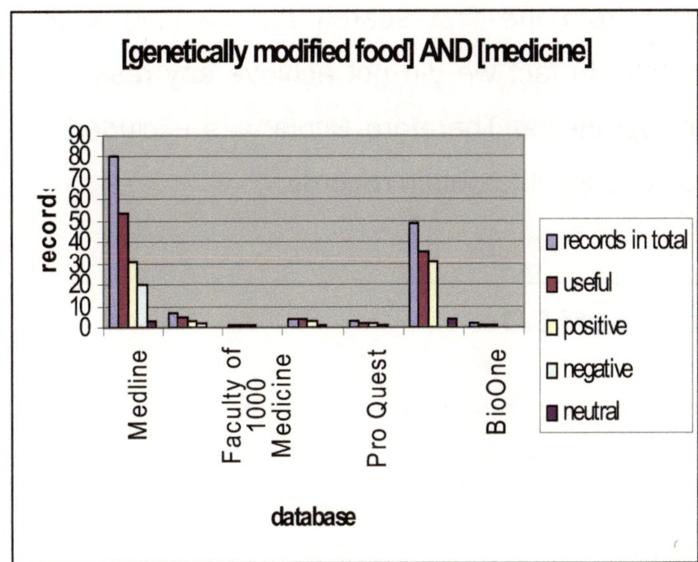

Figure 11. First search in databases.

The records depend only on the intersection of the keywords "genetically modified food" AND "medicine".

Database	records in total	useful	positive	negative	neutral
Medline	80	54	31	20	3
Web of Science	7	5	3	2	0
Faculty of 1000 Medicine	1	1	1	0	0
Faculty of 1000 Biology	4	4	3	1	0
Pro Quest	3	2	2	1	0
Wiley Interscience	49	35	31	0	4
BioOne	2	1	1	0	0

The first search with only the keywords offered us just an amount of 80 records by the Medline database. If we would not have used the Literature Based Discovery Tools, the search would have come now to an end at this point or we had to find connections and co-occuring words by ourselves.

Using the LBD tools presented us a higher amount of records and therefore much more useful papers, we could study to test the hypothesis.

We summarized the achieved records after the second search with the B-terms from first search as C-terms in the following Figure12, p.29.

The B-terms (target concepts) from the first search representing a disease concept had been used for Bitola. In fact we did not achieve any result that we could use for the test of our hypothesis. Therefore Bitola was excluded for the upcoming searches as we did not expect to obtain records.

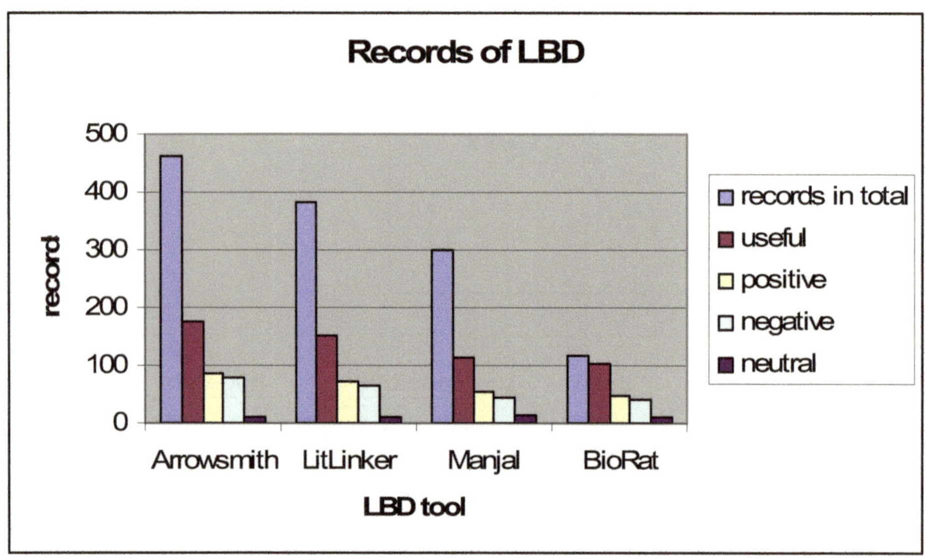

LBD tool	records in total	useful	positive	negative	neutral
Arrowsmith	461	196	97	88	11
LitLinker	382	151	73	64	12
Manjal	299	113	54	44	15
BioRat	561	201	62	111	28

Figure 12. Records obtained by Literature Based Discovery. The records for each C-term (or concept) combined with "genetically modified food" (shown in Fig. 10, p.26) exhibited new B-terms (or target concepts) like presented in Fig.10, p.27. The belonging papers to every B-term (or target concept) had been studied by the scoring rules and then been summarized in this chart.

A search in *Google schola*r was unmanageable due to the record of 52000 links we achieved in the first turn. Thus, we excluded *Google scholar* from subsequent searches as we used the *BioRAT* option searching the *Google* database.

7. Discussion

"No matter how many instances of white swans we may have observed, this does not justify the conclusion that all swans are white."

- Karl Popper[9] -

We can state that science can be defined as a system of knowledge covering general truths or the operation of general laws especially as obtained and tested through scientific method. The nowadays goal of scientists is to contribute new knowledge to science, but in these days researchers do this by the use of the Scientific Method, that can be described as *"principles and procedures for the systematic pursuit of knowledge involving the recognition and formulation of a problem, the collection of data through observation and experiment, and the formulation and testing of hypotheses"* (Merrian-Webster, 2004). The opulence of available digital information, especially in medicine and biochemistry, is such a promising resource that conceptual literature based discovery tools for generating and testing hypothesis are needed in these times of Information explosion.

The methodology of conceptual biology was used to test the hypothesis that genetically modified food has no impact on public health. As the results exhibit we have to deny the hypothesis. With the results obtained by the database and more specific with the tools of Literature Based Discovery we can make the statement, that there is an impact on public health due to the consumption of genetically modified food, because 255 papers show an effect neither positive or negative[10].

The most interesting aspect has been the fact of availability of papers with a negative position. Most of those ones had been found with *BioRAT* in the *Google* database. Unfortunately we can only assume why these thesis did take back seat in the *Google* database, thought they have been published in Journals like *Nature*. The thesis of Arpad Pusztai have been consequently been unavailable in the *Medline* database, instead of the comments according to his papers, that could have been disposable.

---

[9] http://en.wikiquote.org/wiki/Karl_Popper
[10] As databases or LBD tools had papers in common, in fact 255 papers could have been extracted for the study.

The use of *BioRAT* did show some difficulties as the search ran for some keywords more than 30 minutes. Furthermore, the saving of articles and papers was impractical due to the instability of the software. In fact, *BioRAT* did find very help- and useful papers.

We had been very keen to operate with Literature Based Discovery tools using the closed approach, *Manjal, Litlinker* and *Arrowsmith*. There is no denying that *Arrowsmith* is a very helpful, stable and moreover free available tool for the user. With *Arrowsmith* we could make a new connection between two seemingly unrelated fields in science.

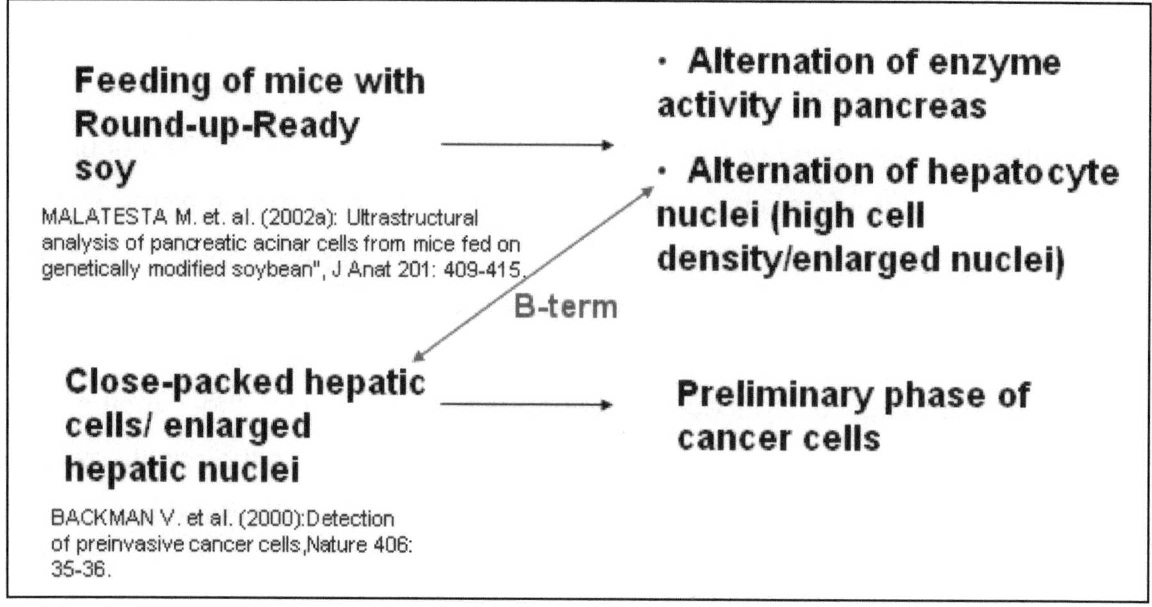

Figure 13. Revealing a new connection between two seemingly unrelated fields in bioscience. With *Arrowsmith* we could obtain a new connection between two papers in different fields of interest, but apparently related by a B-term.

The research group around *Malatesta f*ed mice over 8 months Round-up-Ready soy and analyzed the ultrastructural morphology of pancreatic cells. Furthermore, they discovered an alternation of the hepatic tissue, but they did not a statement where these alternations tend to. *Backman* and his researchers analyzed the reliability and functionality of scanners, detecting cancer and precancer cells. Here the described alternations of *Malatestas`* mice had been classified as precancer cells. This discovery may be emphasized by a study focussing on hepatic precancer in mice by feeding Round-up-Ready soy.

Maybe the sphere of action is restricted as *Arrowsmith* uses only titles and headings of the Medline records, but like Swanson & Smalheiser (1997) found connections during their research on Raynaud and Fish Oil (Swanson 1986) and Migraine and Magnesium (Swanson 1988) we obtained a first new connection. In continuative studies the first list of B-terms could well be scanned for more information or more yet invisible relations between two topics.

To raise another issue, we have to say that *LitLinker* was not a free available software and the results of this study become part of the research group around Wanda Pratt, Assistant Professor of the Information School and the Division of Biomedical & Health Informatics in the School of Medicine at the University of Washington. *LitLinker* and also the *Manjal* software presented redundant UMLS concepts, cases in which a phrase or subphrase matched more than one concept even after disambiguation was attempted. For example, "spinal tap" and "spinal puncture" are two separate UMLS concepts even though they should be a single concept. Another problem is noun-adjective variants. For example, "fibrosis" and "fibrotic" are two separate concepts, as are "necrotic" and "necrosis." From the concept-recognition viewpoint, the adjective is merely a variant of the noun form. Other instances include identical concepts with variations in spelling of the preferred term. Only the last category is recorded in UMLS's ambiguous-terms list. Acronyms resulted in non results when they were present in nonstandard forms that were not recorded in the UMLS. Abbreviations also caused non results when they were identical to non-abbreviated words (like "RAT," which refers to the animal or to Recurrent Acute Tonsillitis).

Using *Manjal* was easy for a newcomer in LBD, as the strong links in such a map did identify pairs of entities that had been well connected in the literature. Displaying the strong connections between the entities had the potential to educate us how to use the new found concepts correctly.
However, the stronger the connection, the less surprising the link had been. In fact, the weaker connections had might actually been more interesting in the sense of exploratory research.

What has been also very suprising was a discrepancy in the intersections found by *Medline* and *Arrowsmith.* In the first run we had 80 records in the database but only 37 intersections by the LBD tool. Unfortunately we cannot state an explanation at this moment.

Finally we can state that LBD tools are powerful and moreover helpful tools in testing a hypothesis, but they should be more readily deployable and accessible. With the information deluge we are facing in biomedical science, it is a challenge for computer scientists to scale up to meet the requirements of this field (Rebholz et al. 2005; Hirschmann et al. 2002). Unlike the information retrieval tools currently available to medical researchers, such as *PubMed,* these tools generate results about possible new connections between medical terms. They also provide an interactive web interface to display the identified correlations in an effective way (Skeels et al.2005). These new discovery systems will help medical researchers capture, and explore new connections in the vast biomedical literature to help them identify new research directions.

8. References

Blagosklonny, M. V. & Pardee, A. B. *Unearthing the gems. Nature* 416, 373 (2002)

Bray, D. *Reasoning for results* Nature 412:863 (2001)

Corney DP, Buxton BF, Langdon WB, Jones DT *BioRAT: extracting biological information from full-length papers*, Bioinformatics.;20(17):3206-13. Epub 2004 Jul 1 (2002)

Cunningham,H., Maynard,D., Bontcheva,K., Tablan,V. *GATE: a framework and graphical development environment for robust NLP tools and applications.* Proceedings of the 40th Anniversary Meeting of the Association for Computational Linguistics (ACL'02), Philadelphia, USA (2002), available at http://www.sims.berkeley.edu/~hearst/irbook/,, 05/08/06

Davies, R. (1989). *The creation of new knowledge by information retrieval and classification.* Journal of Documentation 301-273): 4(45)(1989)

Ganiz, M., Pottenger, W.M., Janneck, C. D. *Recent Advances in Literature based discovery*, dissertation, available at http://wwwlib.umi.com/dissertations/, received 05/03/06

Hirschman, L., Park, J. C., Tsujii, J., Wong, L. & Wu, C. H. *Accomplishments and challenges in literature data mining for biology.* Bioinformatics 18, 1553–1561 (2002).

Hristovski, D., Peterlin, B., Mitchell, J.A., and Humphrey, S.M., *Improving literature based discovery support by genetic knowledge integration*, Stud Health Technol Inform 95: 68-73 (2003)

Lowe, J.B., Baker, C.B, Fillmore, C.J *A Frame-Semantic Approach to Semantic Annotation* In *Proceedings of the ACL SIGLEX workshop ``Tagging Text with Lexical Semantics: Why, What, and How?"*, Washington, D.C., USA (1994)

Merrian-Webster Available at http://www.m-w.com/. received 05/13/06

National Library of Medicine, Link out journals by titles. Available at http://www.nlm.nih.gov/., received 05/13/06

Papadopoulos, Y. *Conceptual Biology-methods and applications,* Internship Report, June 2006

Pratt, W., Yetisgen-Yildiz, M. *LitLinker: Capturing Connections across the Biomedical Literature.* Proceedings of the International Conference on Knowledge Capture (K-CAP'03), Florida (2003), available at http://portal.acm.org/citation.cfm?id=945645.945662&coll=GUIDE&dl=GUIDE&type=series&idx=945645&part=Proceedings&WantType=Proceedings&title=International%20Conference%20On%20Knowledge%20Capture&CFID=69049108&CFTOKEN=95381416, received 05/13/06.

Rebholz-Schuhmann, D. *Facts from text — is text mining ready to deliver.* PLoS Biol. 3, e65 (2005).

Skeels, M., Henning, K., Yetisgen-Yildiz, M., and Pratt, W. *Interaction Design for Literature-Based Discovery.* Proceedings of the International Conference for Human- Computer Interaction (CHI'05), Portland (2005), available at http://www.ischool.washington.edu/wpratt/publications.html, received 05/13/06

Srinivasan, P., *Generating Hypotheses from MEDLINE,* Journal of American Society for Information Science and Technology 55(5): 396-413 (2004).

Swanson, D.R., *Fish oil, Raynaud's syndrome, and undiscovered public knowledge.* Perspectives in Biology and Medicine, 30(1), 7-18 (1986).

Swanson, D. R., *Migraine and Magnesium: Eleven Neglected Connections, Perspect,* Biol. Med. 31: 526-557 (1988).

Swanson, D. R., *Medical Literature as Potential Source of New Knowledge,* Bulletin of Medical Library Association 78(1): 29-37 (1990).

Swanson, D. R., Smalheiser, N. R., *An interactive system for finding complementary literatures: a stimulus to scientific discovery,* Artifical Intelligence 91: 183-203 (1997).

Swanson, D. R. *Online search for logically-related non-interactive medical literatures: A systematic trial-and-error strategy,* Journal of American Society for Information Science 40(5): 356-358(1989).

Weeber, M., Klein, H., and de Jong - van den Berg, L.T.W., *Using Concepts in Literature Based Discovery: Simulating Swanson's Raynaud-Fish Oil and Migraine- Magnesium Examples,* Journal of American Society for Information Science and Technology 52(7): 548-557 (2001).

Weeber, M., Vos, R., Klein, H., de Jong-van den Berg, L.T.W., Aronson, A.R., and Molema G., *Generating Hypotheses by Discovering Implicit Associations in the Literature: A Case Report of a Search for New Potential Therapeutic Uses for Thalidomide,* Journal of the American Medical Informatics Association 10(3): 252-259 (2003).

10. Abbrevations

GATE	General Architecture for Text Engineering
HUGO	Human Genome Organisation
IE	Information Extraction
LBD	Literature Based Discovery
MeSH	Medical Subject Headings
NCBI	National Center for Biotechnology Information
NLM	National Library of Medicine
UMLS	Unified Medical Language System
URL	Uniform Resource Locator
PDF	Portable Document Format
WWW	World Wide Web

11. Acknowledgements

It is important to thank *everyone* that has been a part of such a strange journey.

My first and most important thanks goes to the Lord for setting my family on a path that has yielded a new kind of richness in life, knowledge, and faith.

Furthermore I give severe and lovely thanks to my husband, who was supporting me during each and every midterm and final exam, and during each research milestone. All the time he took up the myriad tasks of family life, graciously "excusing me" to concentrate on my studies. His unconditional support and love during this journey has been a genuine source of inspiration.

In addition, I am more than grateful to my professors Dr. Heinrich Brinck and Dr. Wolfgang Tuma for their support, guidance and their never ending confidence. Without their open-hearted personalities this journey would not have been possible and their intellectual generosity has proved indispensable to completing this internship with sanity and grace.

Moreover, my thanks goes to my exposition "father", to Dr. Barry Commoner. In my life I have met few who possess such a keen intellect and kind spirit. His ability to draw upon, purely from memory, a myriad of complex concepts and then to patiently deposit them upon his students is an ability I hope to someday imitate, but one I never expect to duplicate. I have had the honor and privilege to study under one of the very finest in this professional field.

Not to be forgotten I would like to thank my supervisor Dr. Andreas Athansiou for giving me the golden opportunity to work on the Critical Genome Project and his approval to write this exposition. This project and his immeasurable confidence in my work has provided me with invaluable experience. But more invaluable and priceless is the exchange of ideas, his friendship and his tender care.

More particularly, I would like to thank my friends Miriam Walter, Annette Wolf, Carolin Haubenreich and Yves Texier for supporting me throughout the months of internship as well as for their care and encouragement and moreover for giving me always the positive feedback to the right times. Thank you for giving me the feeling of not being alone in that big town.